Ocean Animals

Sea Stars

by Derek Zobel

BELLWETHER MEDIA
MINNEAPOLIS, MN

Blastoff! Beginners are developed by literacy experts and educators to meet the needs of early readers. These engaging informational texts support young children as they begin reading about their world. Through simple language and high frequency words paired with crisp, colorful photos, Blastoff! Beginners launch young readers into the universe of independent reading.

Sight Words in This Book

a	have	look	their	under
are	here	many	them	water
be	in	on	there	what
called	is	that	they	
can	it	the	this	

This edition first published in 2021 by Bellwether Media, Inc.

Library of Congress Cataloging-in-Publication Data

Names: Zobel, Derek, 1983- author.
Title: Sea stars / by Derek Zobel.
Description: Minneapolis, MN : Bellwether Media, 2021. | Series: Blastoff! beginners: ocean animals | Includes bibliographical references and index. | Audience: Ages PreK-2 | Audience: Grades K-1
Identifiers: LCCN 2020031979 (print) | LCCN 2020031980 (ebook) | ISBN 9781644873977 (library binding) | ISBN 9781648340741 (ebook)
Subjects: LCSH: Starfishes--Juvenile literature.
Classification: LCC QL384.A8 Z63 2021 (print) | LCC QL384.A8 (ebook) | DDC 593.9/3--dc23
LC record available at https://lccn.loc.gov/2020031979
LC ebook record available at https://lccn.loc.gov/2020031980

Editor: Amy McDonald Designer: Andrea Schneider

Printed in the United States of America, North Mankato, MN.

Table of Contents

Sea Stars!

What is that star
in the water?
A sea star!

Sea stars are
ocean animals.
They live in
every ocean.

Sea stars can be many colors. There are many kinds.

giant

sunflower

common

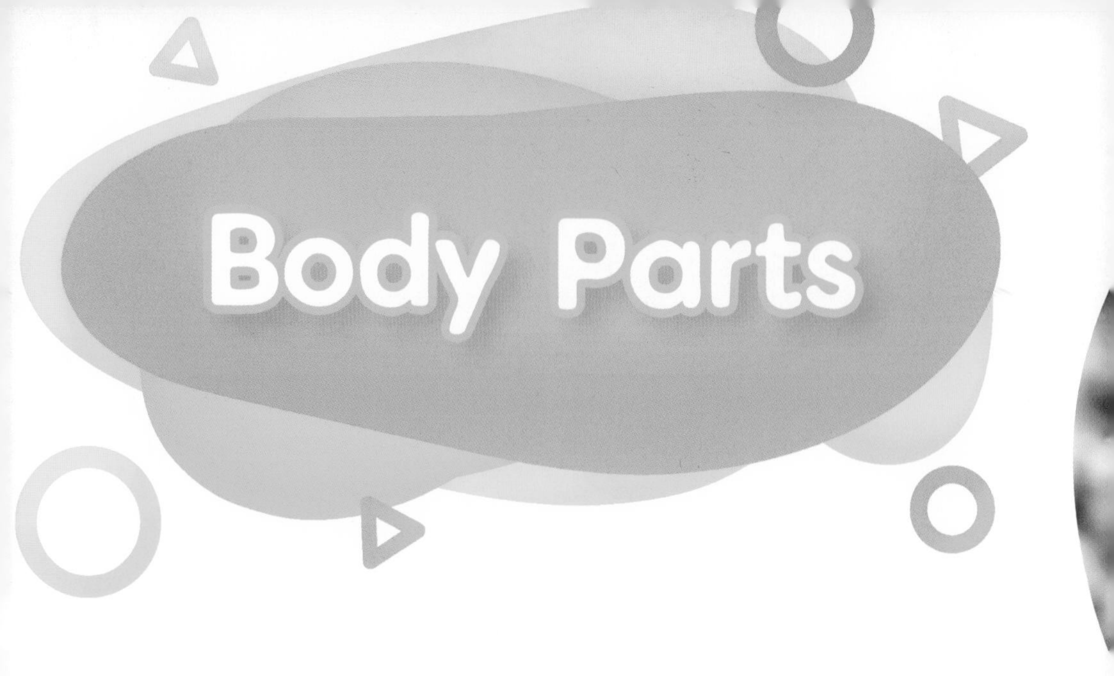

Body Parts

Sea stars
have hard skin.
They are bumpy.

Look under here!
This is the mouth.

mouth

Sea stars have arms. They are called **rays**.

rays

Sea stars have **suckers** on their rays.

suckers

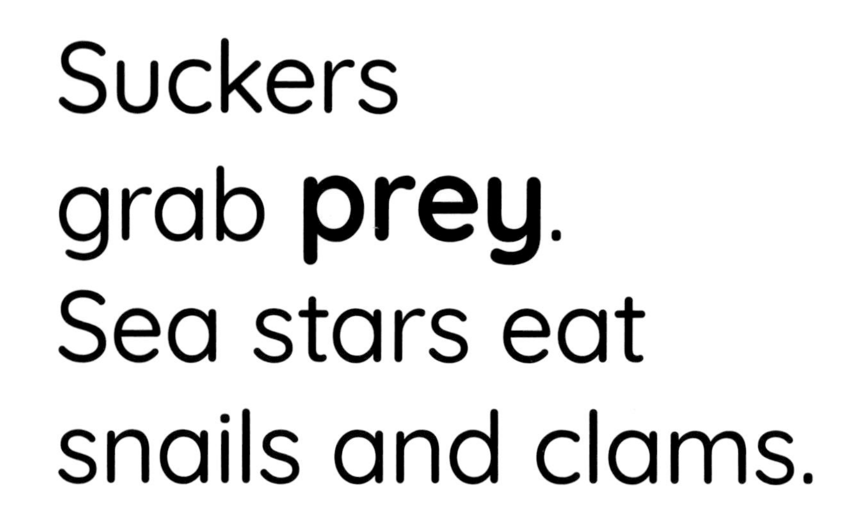

Suckers grab **prey**. Sea stars eat snails and clams.

snail

clam

Lost Rays

This sea star
lost rays.
It grows them
back. Wow!

new rays

Sea Star Facts

Sea Star Body Parts

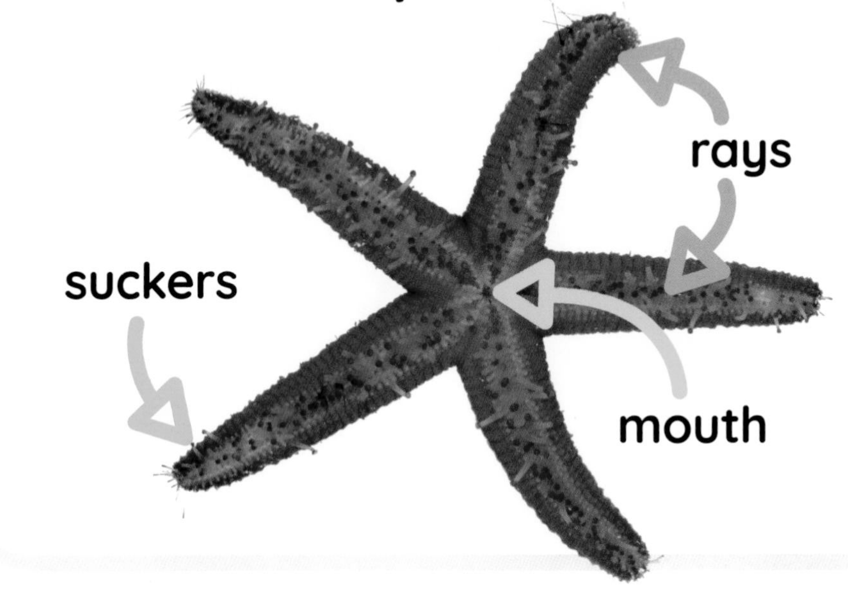

Sea Star Food

Glossary

animals that are food for other animals

the arms of sea stars

body parts that grab and stick

To Learn More

ON THE WEB

FACTSURFER

Factsurfer.com gives you a safe, fun way to find more information.

1. Go to www.factsurfer.com.
2. Enter "sea stars" into the search box and click 🔍.
3. Select your book cover to see a list of related content.

Index

The images in this book are reproduced through the courtesy of: aquapix, front cover; Paper Street Design, p. 3; Vojce, pp. 4-5; Damesea, pp. 6-7, 14; Michael Zeigler, p. 8; Metha Limrosthip, pp. 8-9; Alina Kuptsova, p. 9 (common); Greg Amptman, p. 9 (sunflower); Peter Leahy, p. 10; George Burba, pp. 10-11; Margo Steley/ Alamy, pp. 12-13; Dirk van der Heide, pp. 14-15; Astin Shrader, p. 16; Pascopix/ Alamy, pp. 16-17; Brian Lasenby, p. 18 (snail), 22 (snails); Mark Conlin/ Alamy, pp. 18-19; cbimages/ Alamy, pp. 20-21; Cloudpost, p. 22 (parts); ND700, p. 22 (clams); Dudaev Mikhail, p. 22 (corals); Wiskerke/ Alamy, p. 23 (prey); JaysonPhotography, p. 23 (rays); Dan Bagur, p. 23 (suckers).